Eli Hedley
Beachcomber
Original 1943 Catalog

Reproduced by
BUNGY HEDLEY

Bloomington, IN  Milton Keynes, UK
authorHOUSE™

AuthorHouse™
1663 Liberty Drive, Suite 200
Bloomington, IN 47403
www.authorhouse.com
Phone: 1-800-839-8640

AuthorHouse™ UK Ltd.
500 Avebury Boulevard
Central Milton Keynes, MK9 2BE
www.authorhouse.co.uk
Phone: 08001974150

First published by AuthorHouse 6/23/2006

ISBN: 1-4259-2812-9 (sc)

Printed in the United States of America
Bloomington, Indiana

This book is printed on acid-free paper.

Feel free to drop me a line...

Bungy Hedley
Box 101
Ledbetter, Texas 78946
<bungyh@cvtv.net>

Eli Hedley, Beachcomber, began beachcombing the California, Oregon, and Mexican beaches in 1937. He turned this flotsum and jetsum tossed up from the Pacific Ocean into a million dollar business while living with his wife and four daughters in a tropical paradise right at the sea's edge, called Trade Winds Cove. This cove is now Royal Palms State Park, just up the coast from Los Angeles Harbor.

The unique creations in the catalog were bought by many famous people such as Gary Cooper, Jimmy Stewart, Ronald Reagan, Vincet Price and others. This unusual occupation and Hedley's artistic talent for creating beauty from his beach combings brought him publicity in *MovieTone News* clips, *Wall Street Journal, American Home Magazine, Coronet Magazine, Life Magazine,* and more.

Hedley's art evolved into a tropical and nautical decorating business, and with the resurgent popularity of tropical décor flooding the U.S., Hedley has recently been featured in Tiki Magazine, and is being called a "Tiki Guru" because of many 1950s to 1970s tikis he had hand carved. A recent cataloging of many of his Tikis found them in Disneyland, Reno, Tahoe, Las Vegas, all over the Los Angeles Basin, and even in Tennessee!

This reprint of Eli Hedley, Beachcomber's catalog is dedicated to our dad with much love,
The Hedley Girls

Mare, George Malcome (Mother),
Ba, Bungy, Eli (Daddy), Flo. Circa 1949

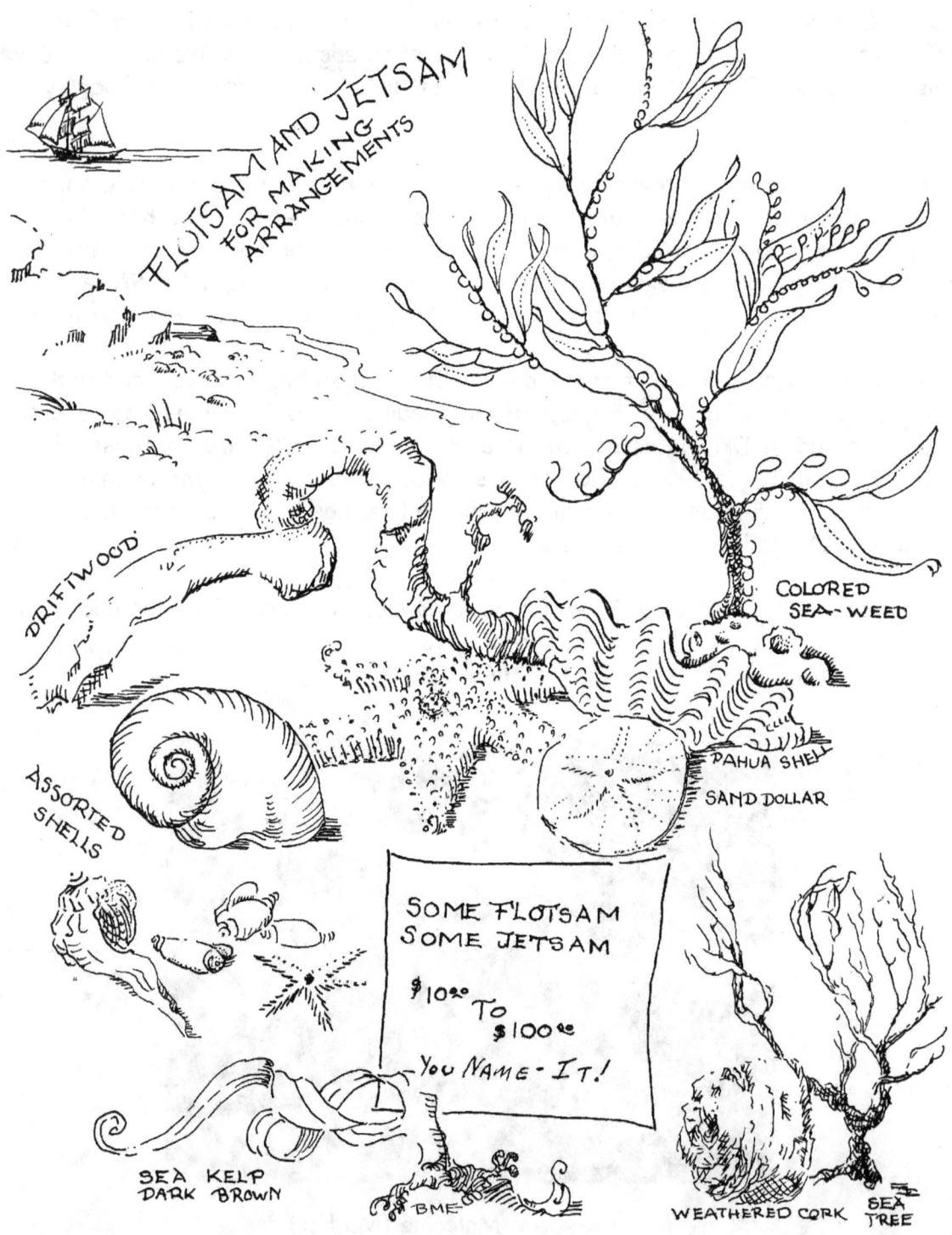

FLOTSAM AND JETSAM
FOR MAKING
ARRANGEMENTS

DRIFTWOOD

COLORED
SEA-WEED

PAHUA SHELL

SAND DOLLAR

ASSORTED
SHELLS

SOME FLOTSAM
SOME JETSAM

$10⁰⁰ TO
$100⁰⁰

YOU NAME - IT!

SEA KELP
DARK BROWN

BME

WEATHERED CORK   SEA TREE

FISH NET — BROWN $.60 SQ.YD.
WEATHERED CORKS $.10 to 35¢

TROPICAL SEAS TABLE $75.00
BRILLIANTLY COLORED
SEA KELP — IN CHEMICAL

Abe Lincoln Desk $45.00
Corner Chair   10.00

4

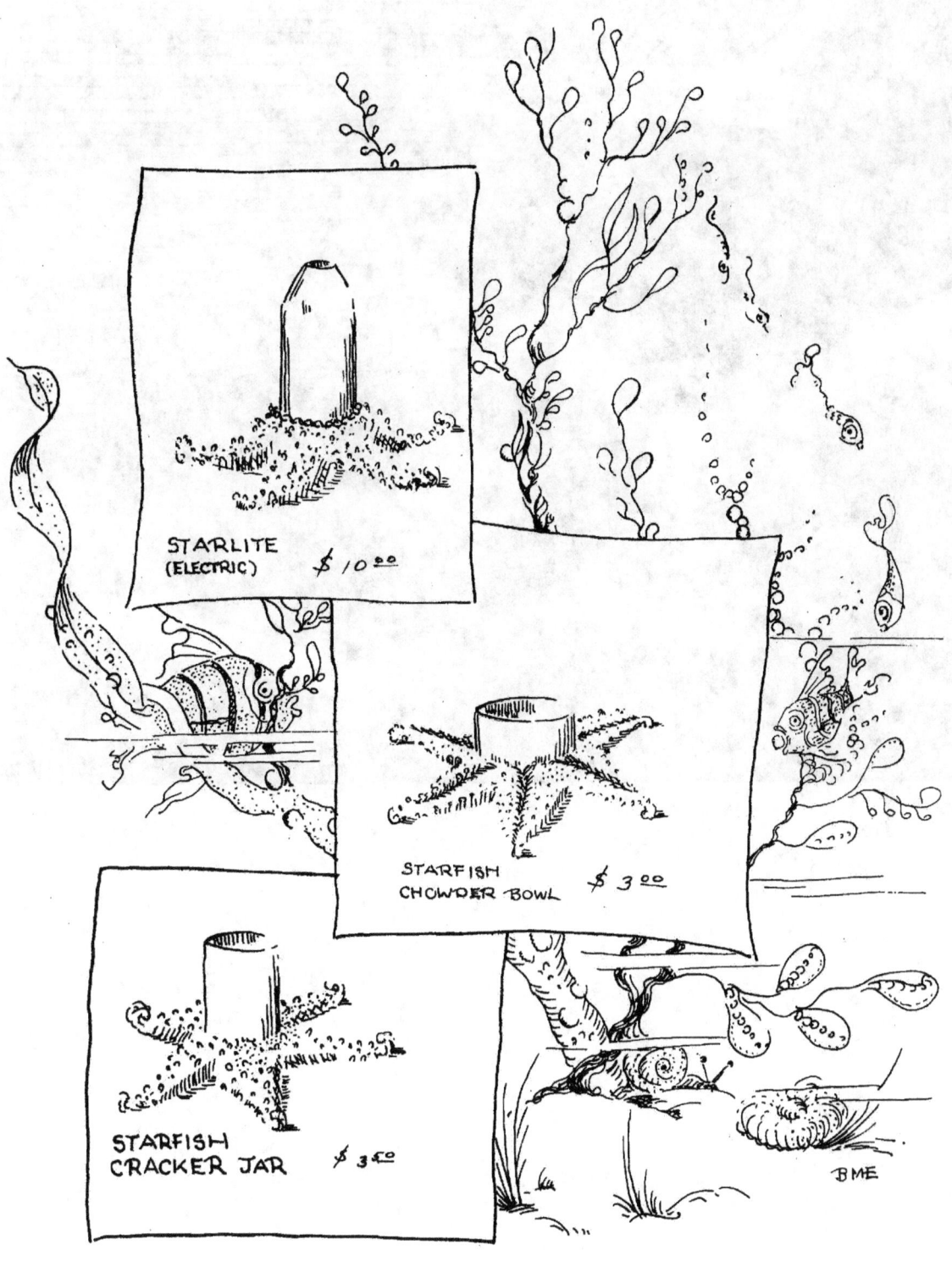

STARLITE
(ELECTRIC) $ 10⁰⁰

STARFISH
CHOWDER BOWL $ 3⁰⁰

STARFISH
CRACKER JAR $ 3⁰⁰

Old Sea Scout Goblet Med. $1.50
Large 2.50

Come and Get It Bell $2.50

STORMY
WEATHER
HURRICANE
TALL $4.50
MEDIUM $2.50

ISLAND HURRICANE
LAMP
LARGE $6.50
SMALL $4.50

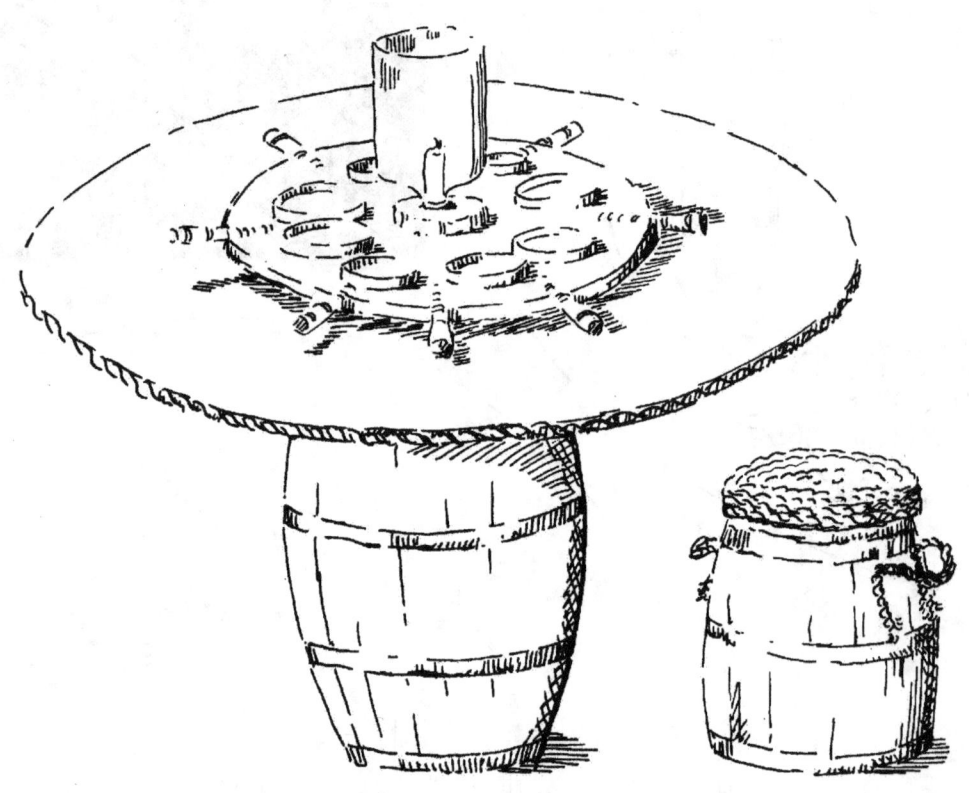

Ship's Wheel Lazy Susan $47.50
ROPED BARREL STOOL ─────── $12.50

Fisherman Candle Lamp $3.50
  "   Electric Lamp $6.50

Beach Comber Candle Lamp $3.50
  "   "   Electric Lamp $6.50

TEEN-AGE
COKE BAR
EQUIPPED $37.50

DRIFTWOOD BENCH $ 35⁰⁰.
RAIN OR SHINE TABLE $27.50
PAK-A-BAIK PONY $ 24⁰⁰

Small Barrell Stave Chair $20.00   Large Barrell Stave Chair $45.00

DARK POLISHED
DRIFTWOOD BENCH $45.00
upholstered with handwoven fish net material
LOW CHUNK SEAT $18.50

coffee
BARREL WELL TABLES $42.50
LOVELY WITH FLOWERS
APPETIZING WITH ICED COKES

MERRY BELL GOBLET $1.50
RING FOR A DRINK

Capt'n Desk $32.50
Corner Chair 10.00

14

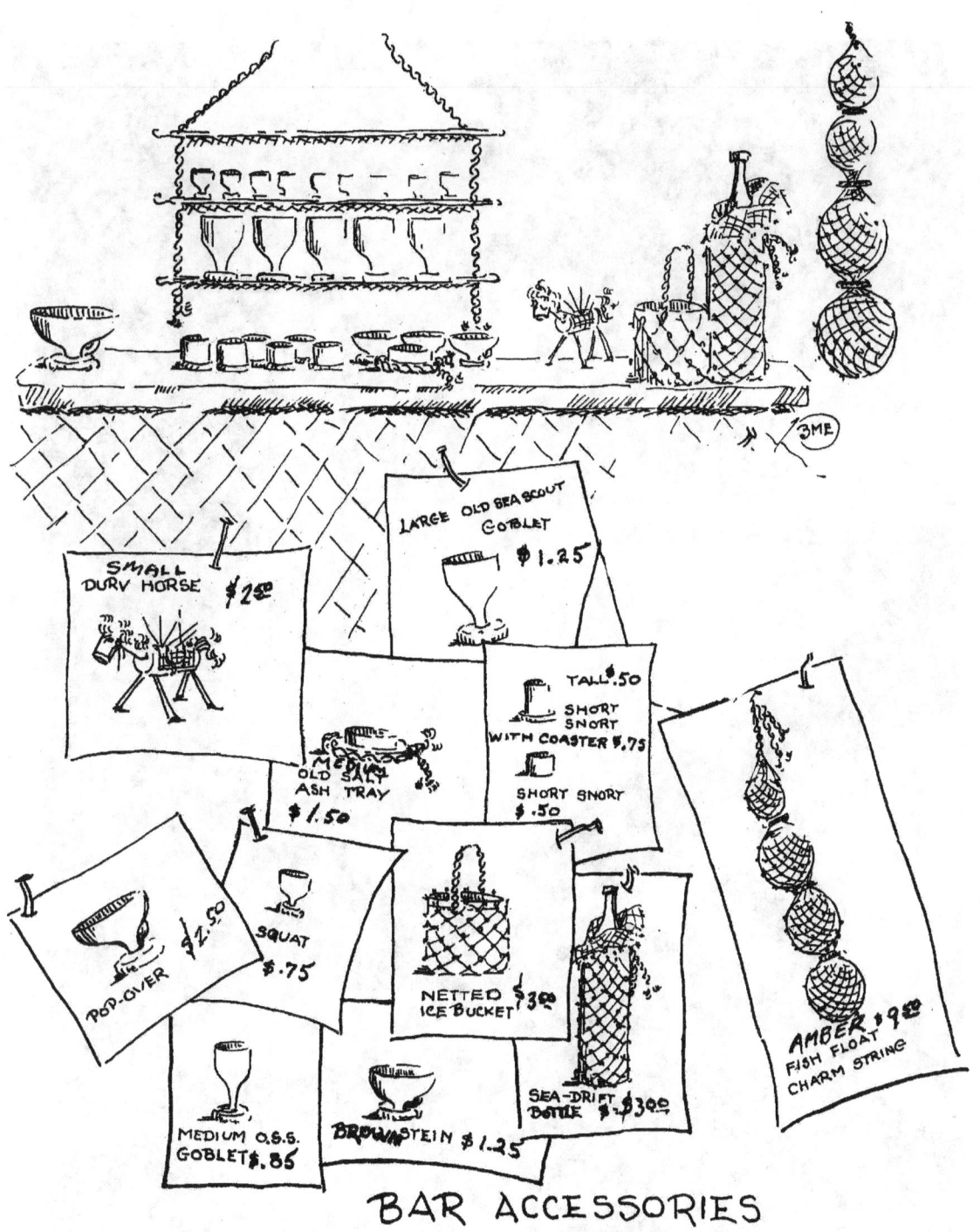

SMALL
DURV HORSE $2⁵⁰

LARGE OLD SEA SCOUT
GOBLET $1.25

TALL $.50
SHORT SNORT
WITH COASTER $.75
SHORT SNORT $.50

MEDIUM
OLD SALT
ASH TRAY $1.50

POP-OVER $2.50

SQUAT $.75

NETTED ICE BUCKET $3⁰⁰

SEA-DRIFT BOTTLE $3⁰⁰

AMBER $9⁰⁰
FISH FLOAT CHARM STRING

MEDIUM O.S.S.
GOBLET $.85

BROWNSTEIN $1.25

BAR ACCESSORIES

15

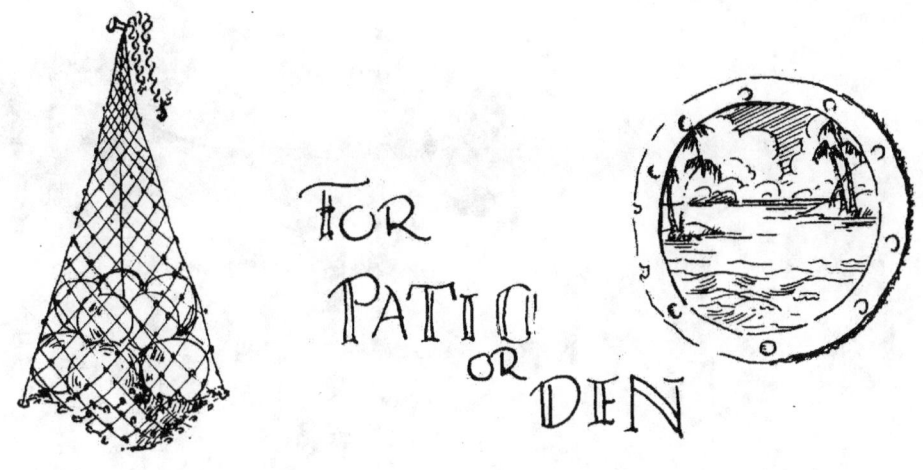

FOR PATIO OR DEN

STARFISH FISH FLOAT CLUSTER LIGHT $ 17.00 (hanging)
BARREL TABLE $ 30.00
DRIFTWOOD ARMCHAIR $ 35.00
DRIFTWOOD MAGAZINE RACK $ 15.00 (stool and magazine rack)
JUG FOOT STOOL $ 10.00 (with roped top plaited)
FISH FLOAT LAMP $ 22.50 (with chart or oil skin shade)
CORK BOOK-ENDS FOR PLANTING $ 3.00
PONY FOR PLANTING SMALL— $ 17.50
WATER COLOR PAINTING, IN PORT HOLE $ 40.00

By Beulah Marker Evans.

Large $4.50

Small $3.50

18

19

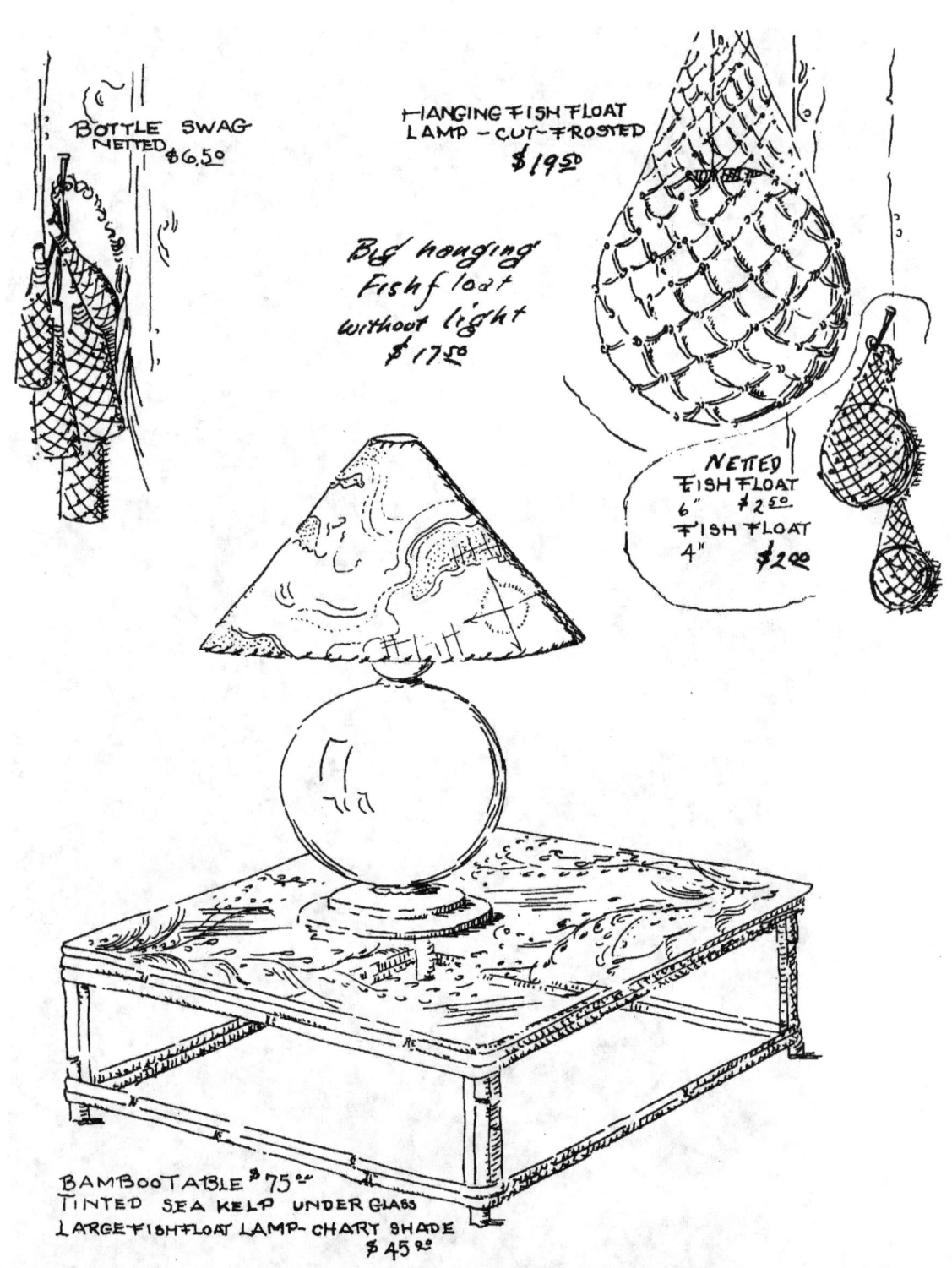

BOTTLE SWAG
NETTED $6.50

HANGING FISH FLOAT
LAMP - CUT-FROSTED
$19.50

Big hanging
Fish float
without light
$17.50

NETTED
FISH FLOAT
6" $2.50
FISH FLOAT
4" $2.00

BAMBOO TABLE $75.00
TINTED SEA KELP UNDER GLASS
LARGE FISH FLOAT LAMP- CHART SHADE
$45.00

Ship Worn Chest $45.00

DRIFTWOOD

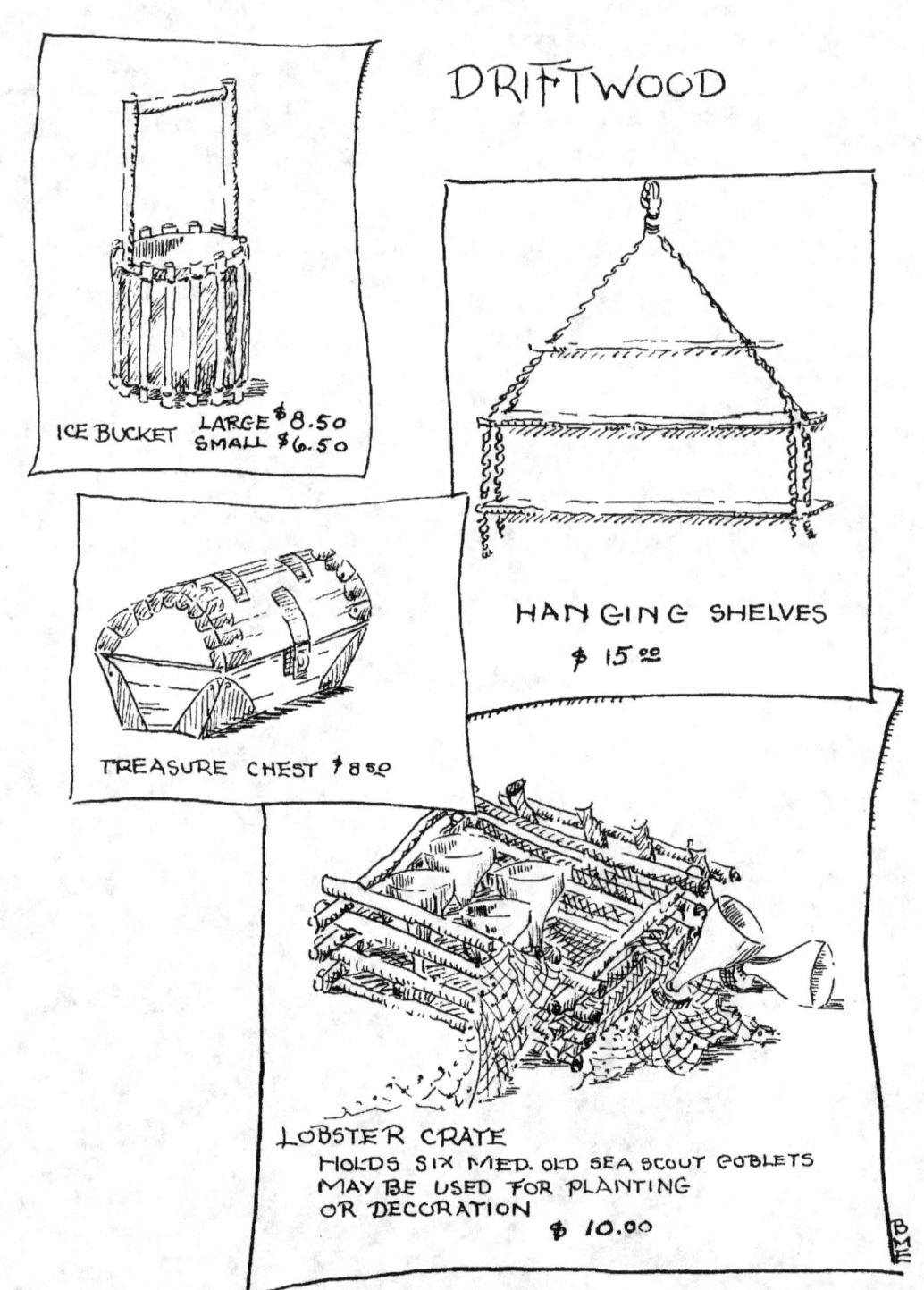

ICE BUCKET LARGE $8.50
            SMALL $6.50

HANGING SHELVES
$ 15.00

TREASURE CHEST $8.50

LOBSTER CRATE
HOLDS SIX MED. OLD SEA SCOUT GOBLETS
MAY BE USED FOR PLANTING
OR DECORATION $ 10.00

ISLAND PIN-UP GIRL

PRICE $6⁵⁰

Addition to Catalog

with family photos
and more of Hedley's creations

Eli Hedley, Beachcomber
Trade Winds Cove

Selection of Hedley's Hand Carved Tikis

Featherstone Tiki, Disneyland 1955
Eli Hedley, Tiki Carver

27

Circa 1956 Hand Carved Tikis

Circa 1956 Hand Carved Tikis
Hedley daughter, Flo Ann

Selection of Hedley's Hard Carved Tikis

HEDLEYS TROPICAL DECORATING JOBS

UNDERWATER SEASCAPE ON MANTLE
HOME ACROSS FROM DISNEYLAND
CIRCA 1956

GIANT CLAM SHELL

TIKIS AT ISLAND TRADE STORE
AND FLO AND BUNGY

Hedley Home
Watercolor by John Fishersmith

Trade Winds Cove Dining Room
with Ship's Wheel Chairs

Living Room at Hedley Family Home
Trade Winds Cove

Trade Winds Cove, 1945
Hills above now filled with homes.

Hedley Home under construction
Trade Winds Cove circa 1946

Trade Winds Trading Company
Hollywood, California circa 1943

Catalog items   1943
Giant hand carved net needle
Driftwood horses
Hurricane lamps

Tiki hut dressing room, tikis, and ba hedley

1943 Life Magazine Article

A DAY'S GLEANING by the Hedleys includes a steer's jawbone (*right foreground*), glass net floats (*left foreground*), a long palm branch (*center*), a bamboo pole, assorted driftwood, a life raft, a buoy (*right background*), a fish trap (*left background*). The whole family poses on the beach with its catch.

BEACHCOMBING BUSINESS

California family makes fortune out of flotsam

One of the weirdest businesses in California, where weird businesses are a perfectly normal thing, is run by a 42-year-old ex-grocer named Weldon Eli Hedley, who makes a living out of the things the Pacific Ocean throws back. As owner-manager of the Trade Winds Trading Company, Hedley pokes up and down a cove on the California coast and salvages driftwood, fish nets, old shoes and whisky bottles. He markets this flotsam in the form of highball glasses, lamps, window drapes and toy horses. Out of his work he expects to take in more than $100,000 in 1946.

A staff of 14 helpers assists Hedley at the Trade Winds Trading Company and, like their boss, work only when they feel like it. Hedley's helpers also include his wife and four daughters, who had faith in him through the meager beginnings of the company. They even stood by when all they had to subsist on was graham crackers and peanut butter. Now that customers all the way from New York to Tahiti are eagerly buying his driftwood, Hedley plans to pack up the whole family and go to the South Sea Islands. He has never been to sea in his life.

TRADE WINDS TRADING COMPANY always has a litter of driftwood in front yard because Hedley claims they never know what they might need.

CONTINUED ON NEXT PAGE

41

Hedley Family Autobiographies:

1. "How Daddy Became A Beachcomber" by daughter Marilyn Hedley. This book is a reprint of the original printed in 1957. It is the story of Eli Hedley, Beachcomber and his family, and how their love of the sea became a way of life and livelihood. Illustrations by Flo Ann Hedley and Marilyn Hedley.

2. "View From the Top of the Mast" by daughter Bungy Hedley. This book is Bungy's view of what an adventure life was, as Eli Hedley's daughter, and her love of sailing, including voyages to Hawaii, Tahiti and the rest of French Oceania, and Mexico. A book filled with unusual experiences.